Spotlight On
Kids Can Code

What Is

MACHINE LEARNING?

Christopher Harris

PowerKiDS
press

Published in 2025 by The Rosen Publishing Group, Inc.
2544 Clinton Street, Buffalo, NY 14224

Copyright © 2025 by The Rosen Publishing Group, Inc.

All rights reserved. No part of this book may be reproduced in any form without permission in writing from the publisher, except by a reviewer.

First Edition

Editor: Greg Roza
Book Design: Michael J. Flynn

Photo Credits: Cover, p. 7 Gorodenkoff/Shutterstock.com; cover, pp. 1, 3–24 (coding background) Lukas Rs/Shutterstock.com; p. 4 SkillUp/Shutterstock.com; p. 5 Summit Art Creations/Shutterstock.com; p. 8 NicoElNino/Shutterstock.com; p. 9 Weerameth Weerachotewong/Shutterstock.com; p. 11 New Africa/Shutterstock.com; p. 13 lilgrapher/Shutterstock.com; p. 15 Zivica Kerkez/Shutterstock.com; p. 17 Pixel-Shot/Shutterstock.com; p. 19 Shutterstock.AI/Shutterstock.com; p. 21 PeopleImages.com - Yuri A/Shutterstock.com; p. 22 IMRAN siddiqi/Shutterstock.com.

Library of Congress Cataloging-in-Publication Data

Names: Harris, Chris G. (Chris Goings), author.
Title: What is machine learning? / Christopher Harris.
Description: [Buffalo, New York] : PowerKids Press, [2025] | Series:
   Spotlight on kids can code | Includes bibliographical references and
   index.
Identifiers: LCCN 2024034358 (print) | LCCN 2024034359 (ebook) | ISBN
   9781499450002 (library binding) | ISBN 9781499449990 (paperback) | ISBN
   9781499450019 (ebook)
Subjects: LCSH: Machine learning–Juvenile literature.
Classification: LCC Q325.5 .H38777 2025 (print) | LCC Q325.5 (ebook) |
   DDC 006.3/1–dc23/eng20240806
LC record available at https://lccn.loc.gov/2024034358
LC ebook record available at https://lccn.loc.gov/2024034359

Manufactured in the United States of America

Some of the images in this book illustrate individuals who are models. The depictions do not imply actual situations or events.

CPSIA Compliance Information: Batch #CWPK25. For Further Information contact Rosen Publishing at 1-800-237-9932.

Contents

Can Machines Really Learn?

We talk about machine learning and **artificial intelligence** (AI), but can computers really learn and think? The answer is yes, but it's different from how you learn and think. Remember, computers are just electrical circuits running **complex** mathematical programs.

Computers rely on the binary system, ones and zeros. In a computer **processor**, a numeral one means there is a flash of electricity, and a zero means there's no electricity; in other words, one is "on" and zero is "off." Human thinking is also based on flashes of electricity between **neurons** in our brains, but we are much more complex than computers.

As you will see in this book, computers can learn through a process we call machine learning, but their learning and thinking is different from how humans learn and think.

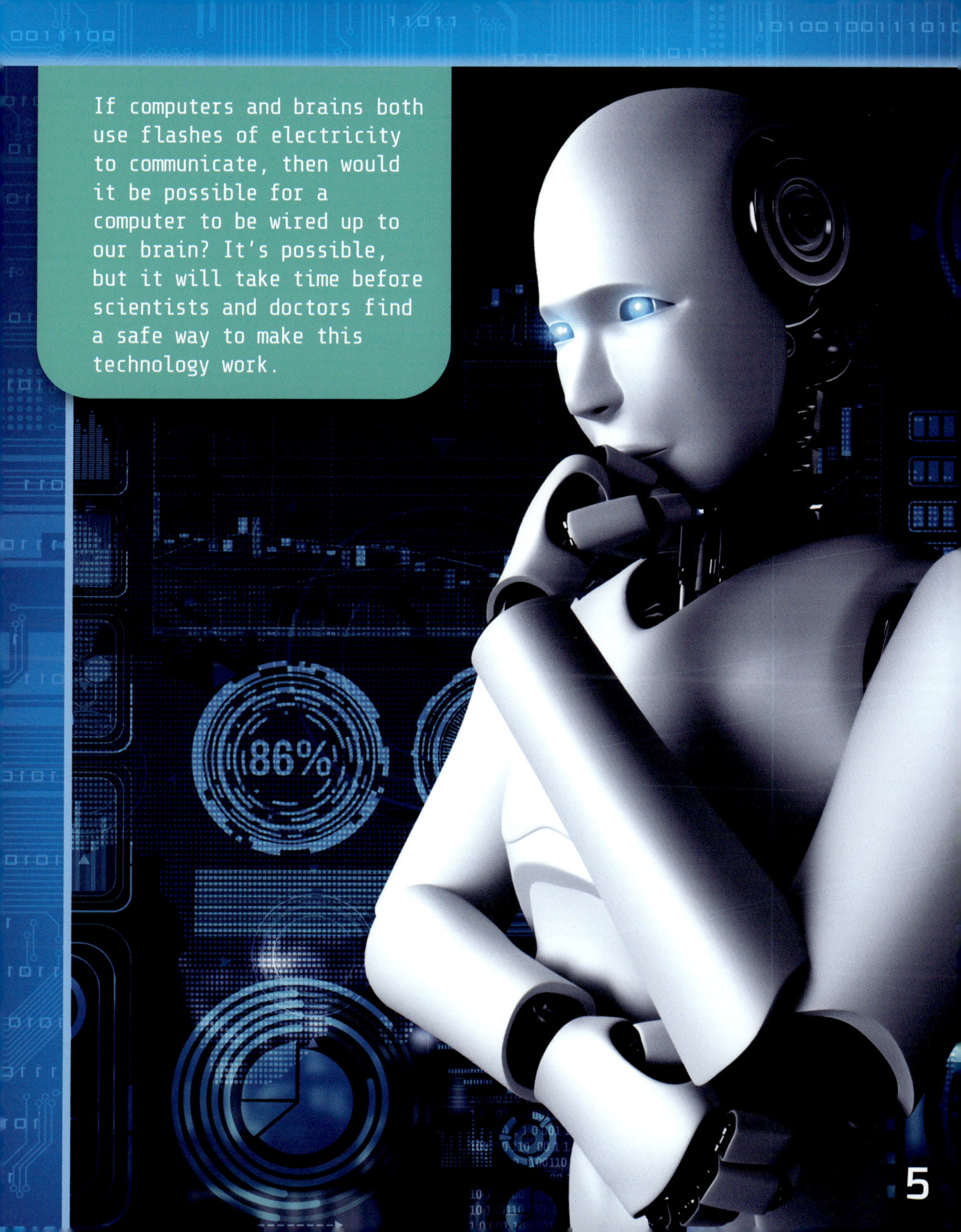
If computers and brains both use flashes of electricity to communicate, then would it be possible for a computer to be wired up to our brain? It's possible, but it will take time before scientists and doctors find a safe way to make this technology work.

86%

Three Types of Machine Learning

There are three main ways to program a computer for machine learning. These are **supervised** learning, unsupervised learning, and **reinforcement** learning. The biggest difference between them is when and how humans are involved in the learning process.

In supervised learning, the most common type of machine learning, a human does most of the work ahead of time by labeling or describing the training data.

On the other hand, in unsupervised learning, humans aren't doing much work at all. The computer is left alone to look at the data and build its own understanding of what it's seeing.

Finally, in reinforcement learning, the human does the work after the computer by providing **feedback** letting the computer know if what it did was right or wrong.

No matter the type of machine learning used, a human still needs to do at least some of the work by creating or labeling a training set of data.

Learning Through Doing

Let's imagine that you want to train a security camera system to tell the difference between a person and a pet. With supervised learning, you provide the computer with lots of different pictures of people and pets. Your job is to label the pictures as a person or a pet before loading them into the computer. This will help the computer figure out the differences between the two sets of pictures so it can learn to identify the two types.

With unsupervised learning, the computer is given all the pictures, and it has to find the differences and create its own groups and labels. Maybe the computer learns that people are tall, and pets are short. It might learn that pets have tails, and people don't. Or maybe it looks at the number of legs on pets and people.

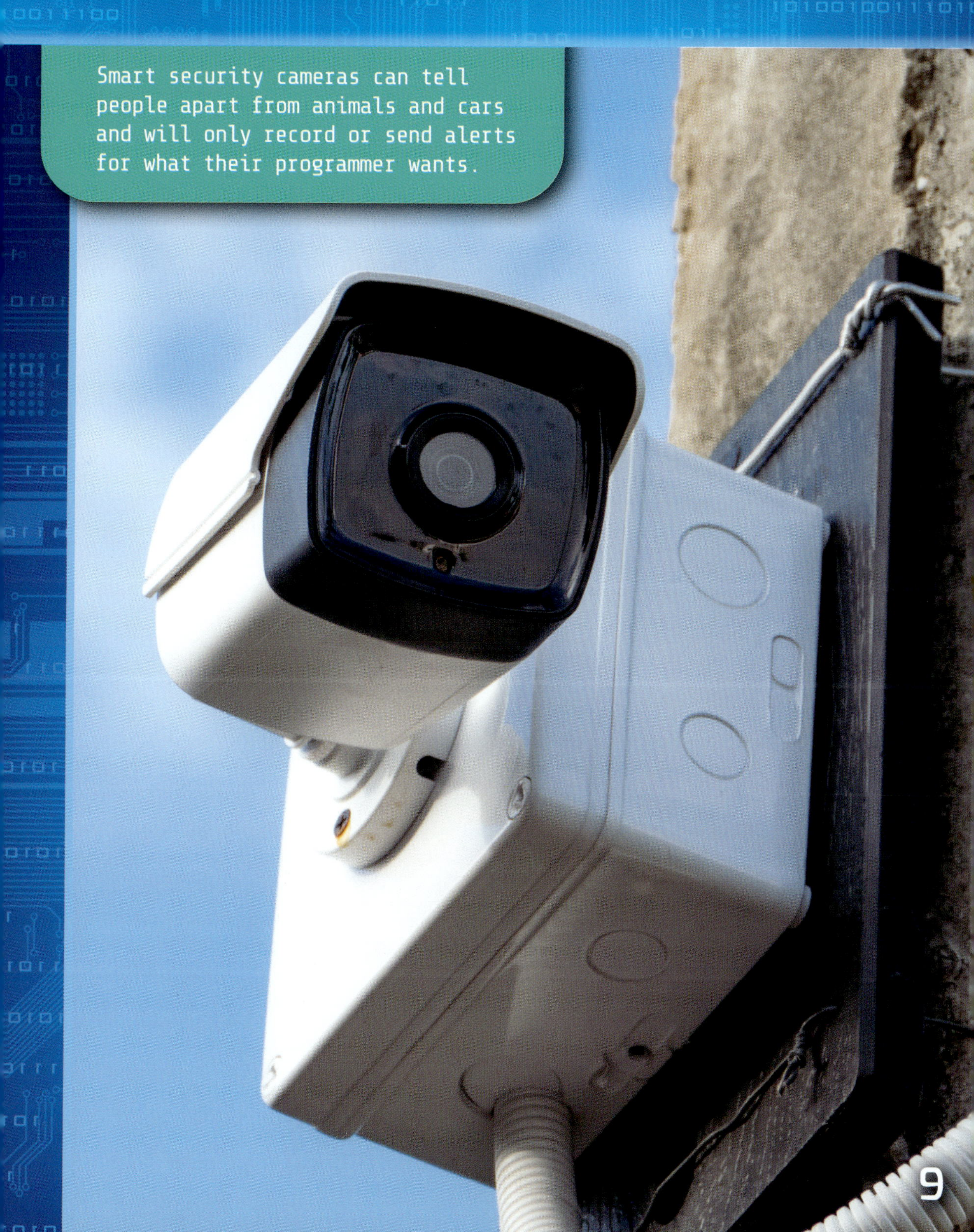
Smart security cameras can tell
people apart from animals and cars
and will only record or send alerts
for what their programmer wants.

Learning Through Feedback

To use reinforcement learning for the security camera example, you could train a program to send an alert only for detecting people. If the camera sends an alert every time it sees movement, you could give it points if the alert is for a person and take points away if it's a pet.

Over time, the computer will learn based on the reinforcement—positive points for correctly recognizing a person and negative points for calling a pet a person. To get points, the computer will have to figure out what a person looks like so it only sends positive alerts.

The result is similar to the other types of machine learning, but your time and work as the human is different.

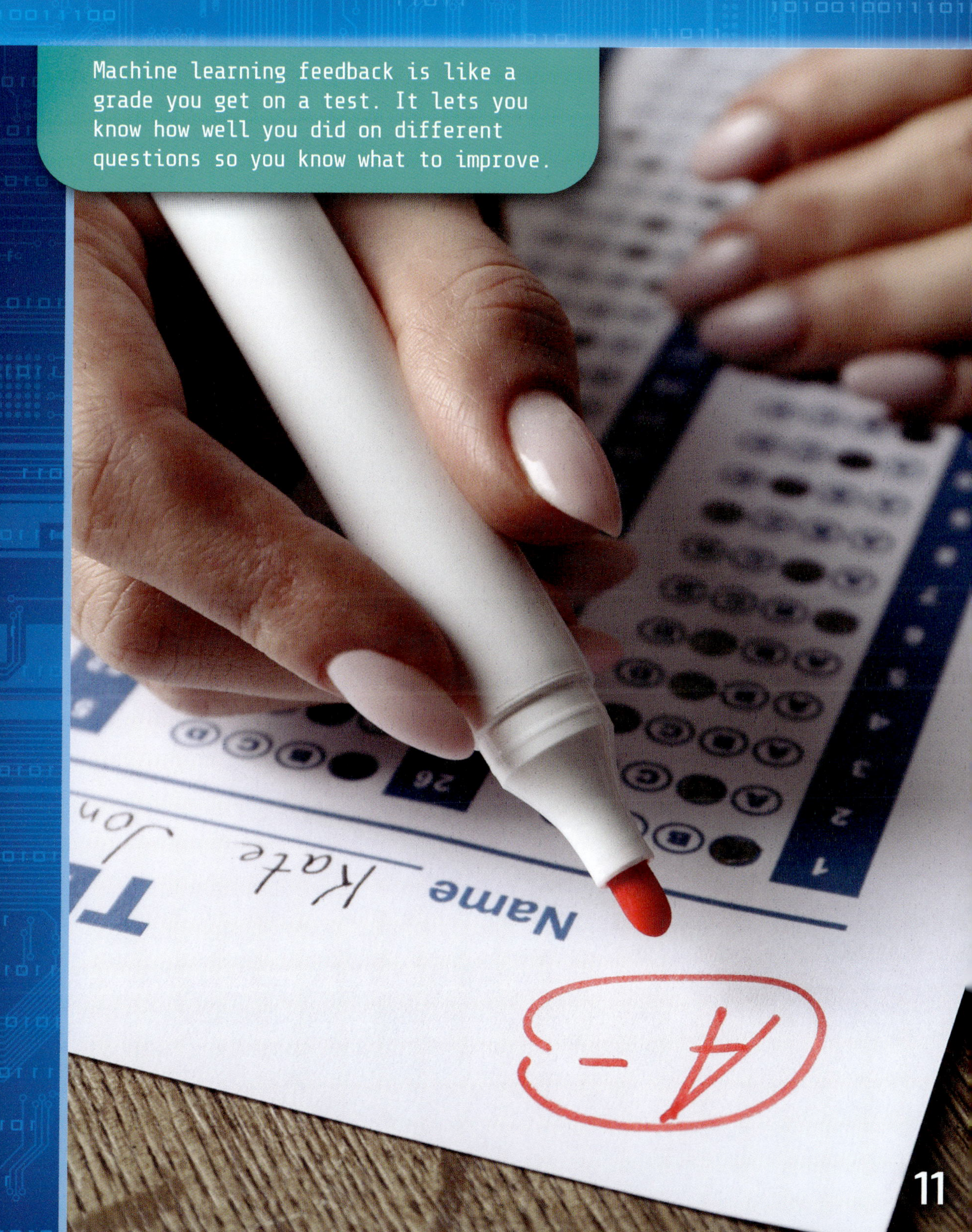

Machine learning feedback is like a grade you get on a test. It lets you know how well you did on different questions so you know what to improve.

Every Click Is a Lesson

The human work of machine learning can be shared by a group of people. For example, every time a website asks you to click on pictures of bicycles to prove you aren't a robot, that is also an example of reinforcement learning. You are helping a machine learn what a bicycle looks like by either labeling a set of images for supervised learning or providing feedback on reinforcement learning.

If you wanted to speed up the training for your security camera program, you could use the other students in your class as helpers. You could set up a touch screen at the classroom door that shows pictures and ask your fellow students to tap on the pictures that include a pet. With every tap, your camera program would learn more.

Breaking the Code

CAPTCHA tests (automated tests to tell computers and humans apart) are also a way to get many people to help with reinforcement learning. Older CAPTCHA tests asked users to type in words scanned from books. Now most CAPTCHA tests are based on images as the focus of machine learning.

Every time you fill out a CAPTCHA test,
you may also be doing work for technology
companies. They are using you, as a human,
to help with machine learning.

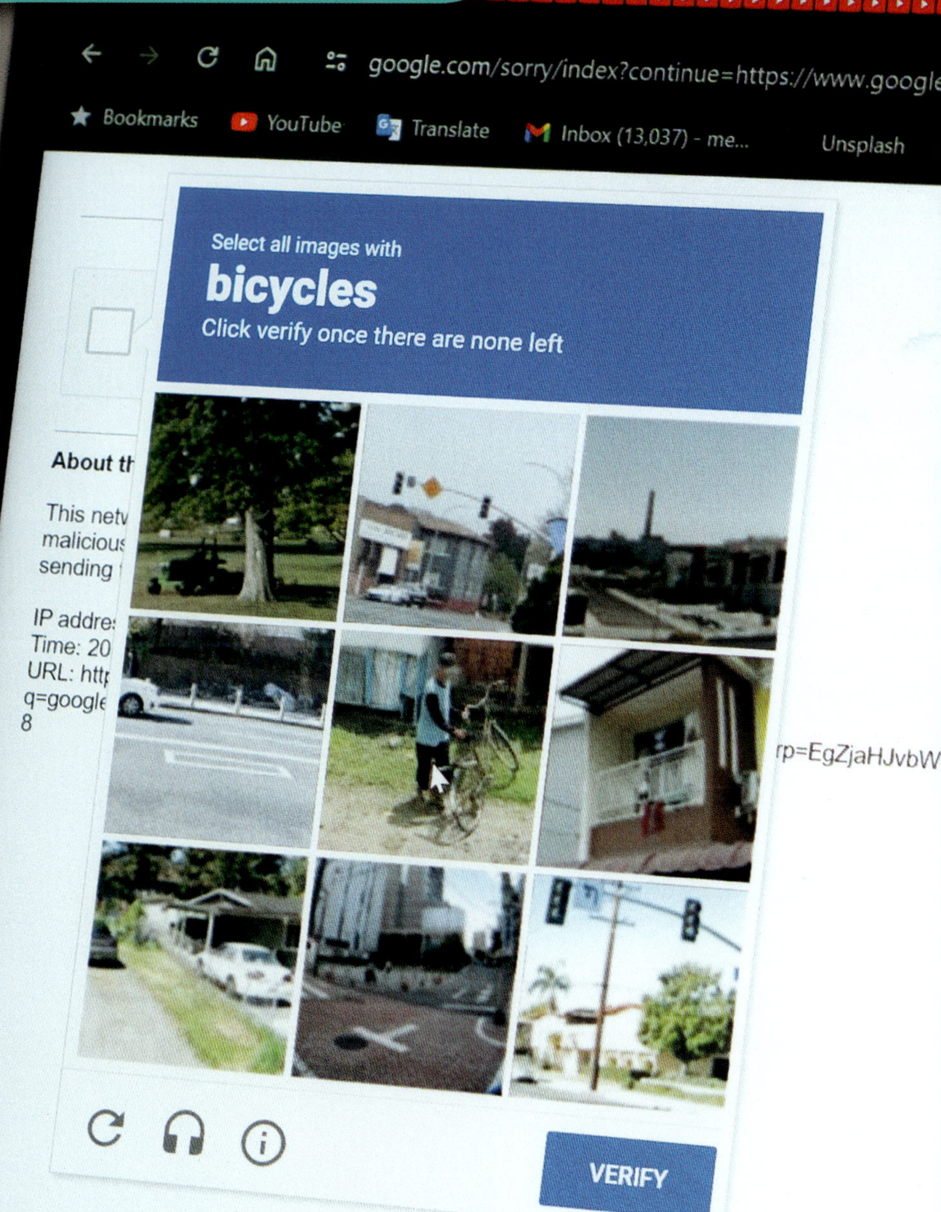

google.com/sorry/index?continue=https://www.google.
Bookmarks
YouTube
Translate
Inbox (13,037) - me...
Unsplash
Select all images with
bicycles
Click verify once there are none left
About th
This netw
malicious
sending
IP addres
Time: 20
URL: http
q=google
8
rp=EgZjaHJvbWU
VERIFY

Finding Patterns

What computers are really doing when they go through machine learning is identifying patterns. Human brains love patterns, and that's how we learn as well.

If you have a younger sibling, maybe you thought it was funny if they pointed at a cat and said "dog." This shows a problem in a pattern. Both cats and dogs have four legs, fur, and tails, so it's easy to understand their confusion. When someone corrects them and says "No, that's a cat," then the child gains a better understanding of the pattern.

They might learn that cats purr and meow instead of barking, or they might realize that cats often have pointed ears. Some dogs have pointed ears as well, though, so this can create new problems in the pattern.

How many similarities and differences between cats and dogs can you identify? The more information you can provide, the better the computer can learn through testing examples.

Learning from Big Data Sets

The best way to learn about all the differences between the many types of cats and dogs is to look at pictures of thousands of examples of as many types as possible. This could be millions of pictures. In computing, this is known as a big data set because it is so large that our human brains struggle to work with the data.

Computers, however, are great at understanding big data sets, and this is one of the reasons machine learning works. If you give a computer enough data, it can find patterns. For the best machine learning, though, you need very big sets of data. Without enough data, the computer won't be able to identify a pattern or might make mistakes in its learning.

Breaking the Code

How big is a big data set? ChatGPT-4 was trained on about 10 trillion words. A typical elementary school library may have about 12,000 books, which contain about 6 million words. That means it would take more than 1.5 million school libraries worth of books to equal the size of the training set for ChatGPT-4.

Computers can organize and understand far more information than humans. For example, a computer can learn every aspect of every type of dog.

Building Models for Understanding

With these enormous sets of data, machine learning has contributed to the fast development of generative artificial intelligence. The large language models (LLMs) that power ChatGPT have been trained on almost all the books, newspapers, and magazines that have ever been written and digitized. The LLM has so many examples of writing that it has been trained to generate new writing that may be similar to what a human would create.

LLMs have also been trained using artwork and photographs so they can generate new images. The basic idea is the use of an enormous set of examples combined with machine learning to find patterns and build a model for the computer to create new output that matches the original examples.

With an LLM that includes different art styles from classical painters, you can ask an AI to generate a new image of a dog in the style of impressionist painter Monet.

The Role of Machine Learning in AI

Even though artificial intelligence seems to work without a human involved, AI is usually based on humans supervising or reinforcing machine learning. The P in ChatGPT stands for "pre-trained." These new **versions** of AI have been more successful because they were better trained using larger sets of examples.

There have been problems when AI didn't have enough examples for good machine learning. For example, smartphone camera programs were mostly trained with pictures of people with lighter skin, so they struggled to take pictures of people with darker skin. The phone makers had to use different sets of training pictures to change the machine learning in the camera phone AI. Creating better AI that works for everyone must start with better machine learning.

Breaking the Code

Dr. Joy Buolamwini helped form the **Algorithmic** Justice League to fight back against **bias** in AI. As a woman of color, she struggled to get **facial recognition** programs to see her as a person because the training sets didn't include faces that looked like hers. Her work has helped raise awareness of these problems.

Concerns about facial recognition go beyond bias for skin tone. People also worry about losing privacy when public cameras use facial recognition to track people.

The Future of Machine Learning

Our world is becoming increasingly focused on meeting individual wants and needs. We expect song recommendations and an endless feed of new videos that will make us laugh. All of that requires machine learning to create models that can power the AI suggestion algorithms.

With the importance that machine learning has on training artificial intelligence programs, we need all types of people to study and work in the field. Think about the camera program trained to tell the difference between cats and dogs. What about people that have ferrets as pets? If their ferret escaped, they would want a camera to track it as a pet as well! That will only happen if someone helping with the machine learning for the camera program thinks about all the possible kinds of pets and makes sure the computer learns about them.

Glossary

algorithm: A set of steps that are followed in order to solve a mathematical problem or complete a computer process.

artificial intelligence: The ability of some computers and programs to solve problems the way people do.

bias: A way of thinking that always favors one way of feeling or acting over any other. In AI, this means results produced by algorithms that are unfair or distorted from reality.

complex: Not easy to understand or explain; having many parts.

facial recognition: A way for computers to use technology to identify a person's face.

feedback: A response that suggests how to change or improve something.

neuron: A cell that is the basic working unit of the nervous system and that carries nerve impulses.

processor: A part of a computer that processes (sorts out) data.

reinforcement: A response to a computer's output intended to make future output more accurate.

supervise: To watch over or guide.

version: A form of something that is different from the ones that came before it.

Index